Theoretical Foundations

of Structure-Behavior

Coalescence

William S. Chao

Structure-Behavior Coalescence

Systems Architecture **=** **Systems Structure** **+** **Systems Behavior**

CONTENTS

6

ABOUT THE AUTHOR

Dr. William S. Chao is the CEO & founder of SBC Architecture International®. SBC (Structure-Behavior Coalescence) architecture is a systems architecture which demands the integration of systems structure and systems behavior of a system. SBC architecture applies to hardware architecture, software architecture, enterprise architecture, knowledge architecture and thinking architecture. The core theme of SBC architecture is: Architecture = Structure + Behavior.

William S. Chao received his bachelor degree (1976) in telecommunication engineering and master degree (1981) in information engineering, both from the National Chiao-Tung University, Taiwan. From 1976 till 1983, he worked as an engineer at Chung-Hwa Telecommunication Company, Taiwan.

William S. Chao received his master degree (1985) in information science and Ph.D. degree (1988) in information science, both from the University of Alabama at Birmingham, USA. From 1988 till 1991, he worked as a computer scientist at GE Research and Development Center, Schenectady, New York, USA.

Dr. William S. Chao has been teaching at National Sun Yat-

Sen University, Taiwan since 1992 and now serves as the president of Association of Enterprise Architects, Taiwan Chapter. His research covers: systems architecture, hardware architecture, software architecture, enterprise architecture, knowledge architecture and thinking architecture.

PART I: FORMAL DESCRIPTION OF THE SYSTEMS ARCHITECTURE

Structure-Behavior Coalescence Means to Integrate the Systems Structure and Systems Behavior

Systems structure and systems behavior are the two most prominent views of a system, integrating the systems structure and systems behavior is apparently the best way to achieve an integrated whole of a system.

If we are not able to integrate the systems structure and systems behavior, then there is no way that we are able to integrate the whole system.

Structure-behavior coalescence (SBC) provides an elegant way to integrate the systems structure and systems behavior of a system. In other words, SBC facilitates an integrated whole of a system.

Interaction

An interaction represents an indivisible and instantaneous handshake or communication between two agents. The caller agent (either external environment's actor or component) communicates with the callee agent (component) through the interaction.

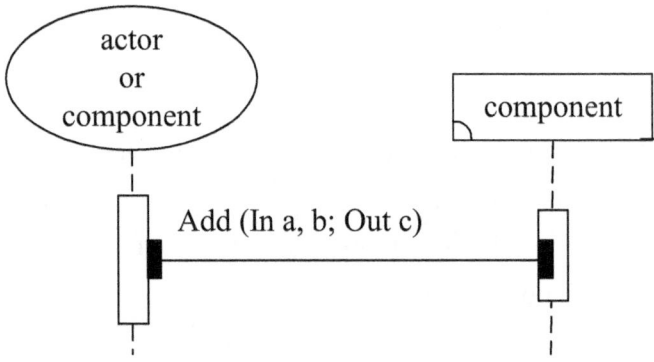

There are two ports, i.e., calling port or called port, of an interaction. The caller agent owns the "calling port" of the interaction.

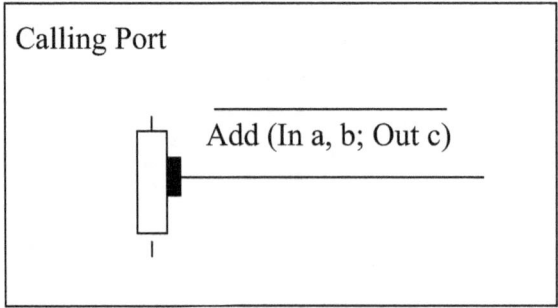

The caller agent together with the "calling port" is named the "calling action".

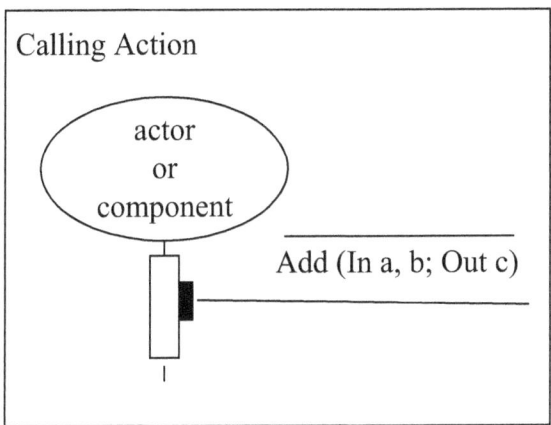

The callee agent owns the "called port" of the interaction.

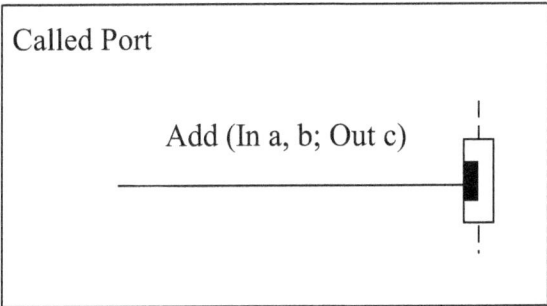

The callee agent together with the "called port" is named the "called action".

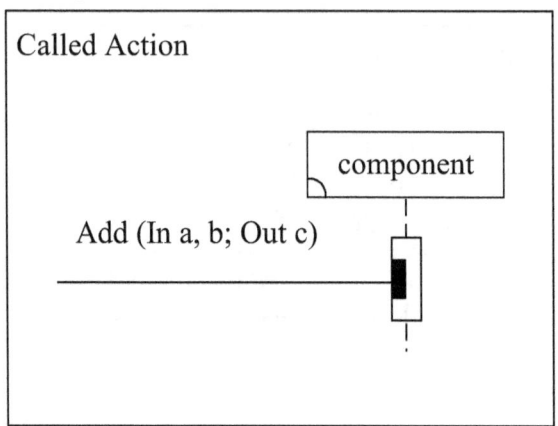

In order to simplify the interaction diagram, we will redraw it as follows.

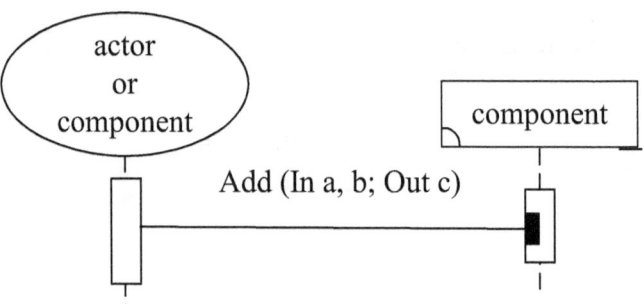

Interactions among Components and Actors to Draw Forth the Systems Behavior

In a system, if the components, and among them and the external environment's actors to interact (or handshake), these interactions will draw forth the systems behavior.

We conclude that "interaction" plays an important factor in integrating the systems structure and systems behavior for a system.

The overall behavior of a system consists of many individual behaviors. Each individual behavior represents an execution path. We use an interaction flow diagram (IFD) to demonstrate this individual behavior.

Collection of All Interaction Flow Diagrams Defines the Systems Architecture

The collection of all interaction flow diagrams defines the integration of systems structure and systems behavior of a system.

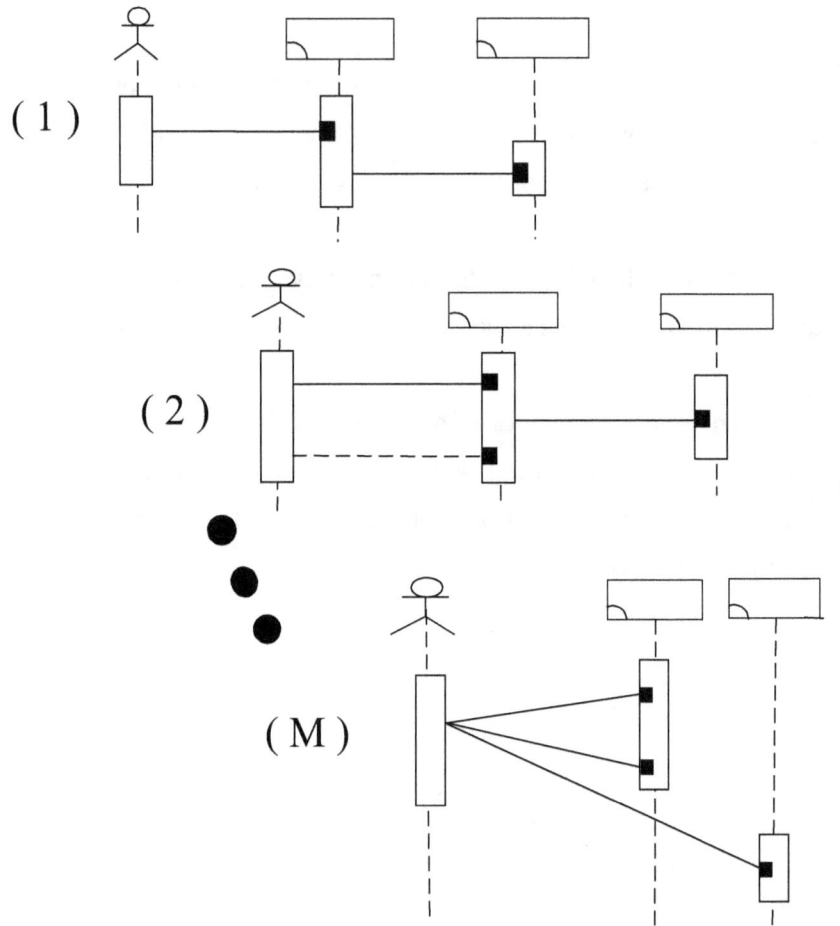

That is, the collection of all interaction flow diagrams defines the systems architecture.

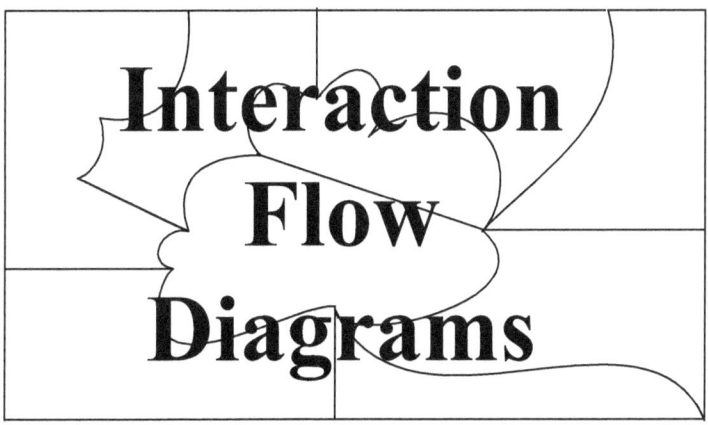

Systems architecture (SA) represents a knowledge repository of a system. Stakeholders can submit and acquire knowledge to and from this knowledge repository.

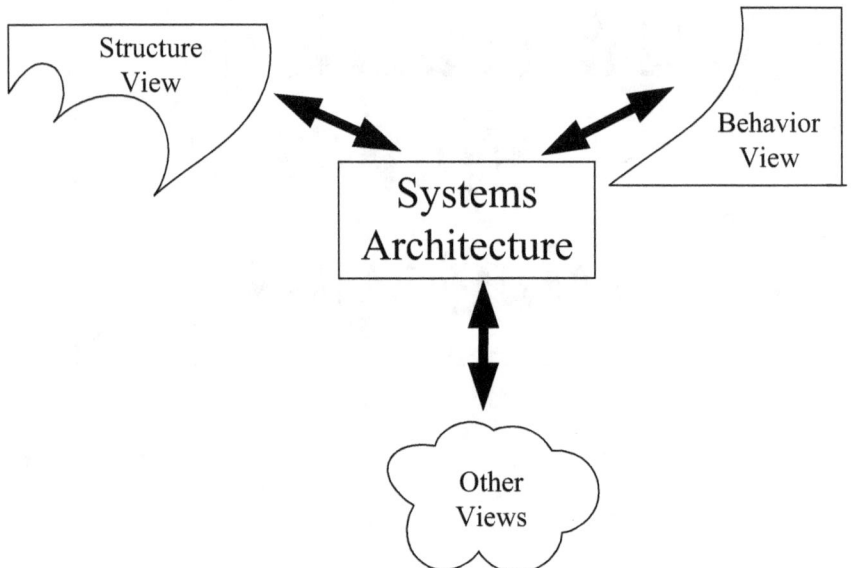

So, the collection of all interaction flow diagrams represents a knowledge repository of a system. Stakeholders can submit and acquire knowledge to and from this knowledge repository.

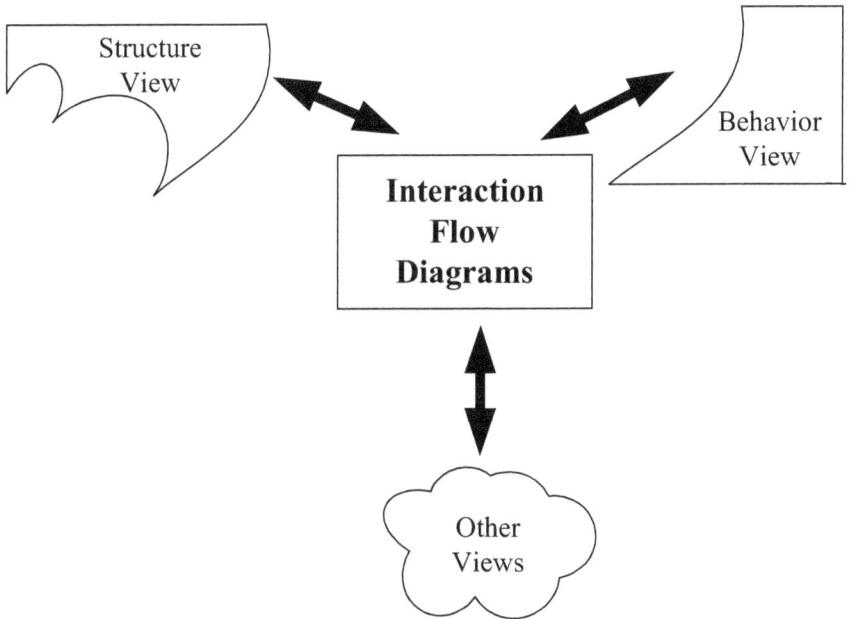

Formal Description of an Interaction

We formally describe an operation-based interaction as a 5-tuple INTERACTION = <operation_call_or_return, caller_actor_or_component, operation_name, i/o_parameters, callee_component>, where "operation_call_or_return" stands for an OPERATION_CALL or OPERATION_RETURN tag, "caller_actor_or_component" stands for the name of a caller actor or component, "operation_name" stands for the name of an operation, "i/o_parameters" stands for a 2-tuple of <input_parameters, output_parameters> where "input_parameters" stands for a set of input parameters and "output_parameters" stands for a set of output parameters, and "callee_component" stands for the name of a callee component.

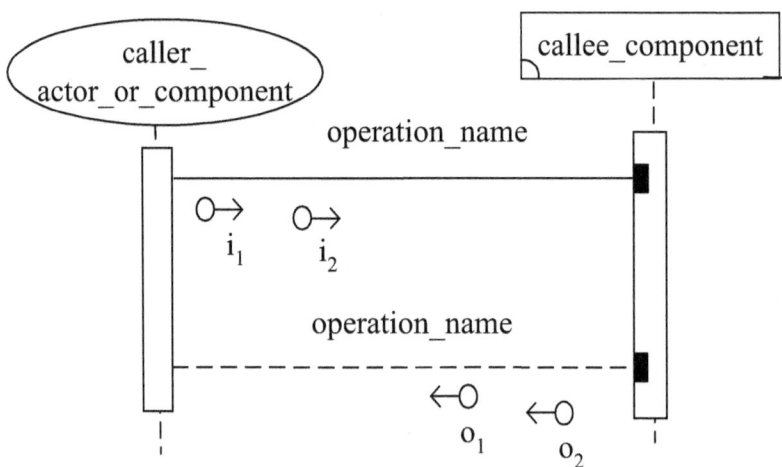

Formal Description of an Interaction Flow Diagram

An interaction flow diagram (IFD) is constructed by a sequence of interactions among the components and external environment's actors.

We formally describe the xth interaction flow diagram (IFD) of the systems architecture as a sequence $IFD_x = (interaction_{xz})_{z = 1 \text{ to } N}$, where "z" stands for the zth (z = 1 to N) interaction of this xth interaction flow diagram.

Or we can formally describe the xth interaction flow diagram as a sequence $IFD_x = (interaction_{x1}, interaction_{x2}, interaction_{x3}, ..., interaction_{xN})$.

Examples of Formal Description of an Interaction Flow Diagram

We suppose the example systems architecture consists of two interaction flow diagrams.

The first interaction flow diagram consists of two interactions.

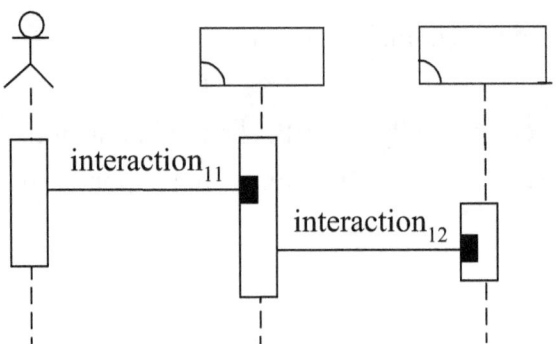

We formally describe this 1st interaction flow diagram as a sequence $IFD_1 = (interaction_{11}, interaction_{12})$.

The second interaction flow diagram consists of three interactions.

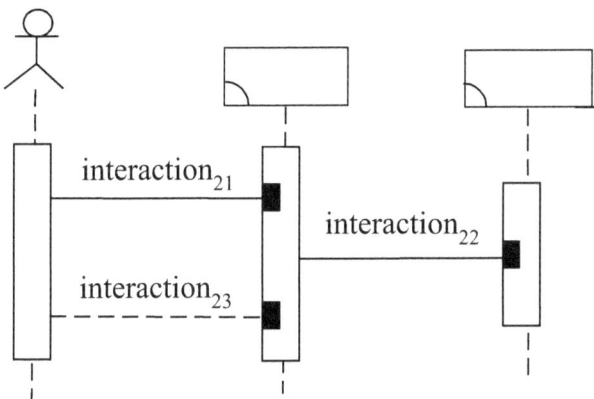

We formally describe this 2nd interaction flow diagram as a sequence $IFD_2 = (interaction_{21}, interaction_{22}, interaction_{23})$.

Formal Description of the Systems Architecture

Since the collection of all interaction flow diagrams defines the systems architecture, the systems architecture can be formally described as a M-tuple IFDTUPLE = $<IFD_x>_{x = 1 \text{ to } M}$, where "x" stands for the xth interaction flow diagram of this collection of all interaction flow diagrams.

Or we can formally describe the systems architecture as a M-tuple IFDTUPLE = $<IFD_1, IFD_2, IFD_3, ..., IFD_M>$.

Examples of Formal Description of the Systems Architecture

We suppose the example systems architecture consists of two interaction flow diagrams.

The first interaction flow diagram consists of two interactions.

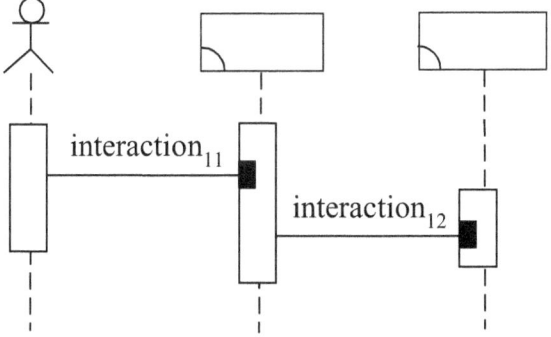

The second interaction flow diagram consists of three interactions.

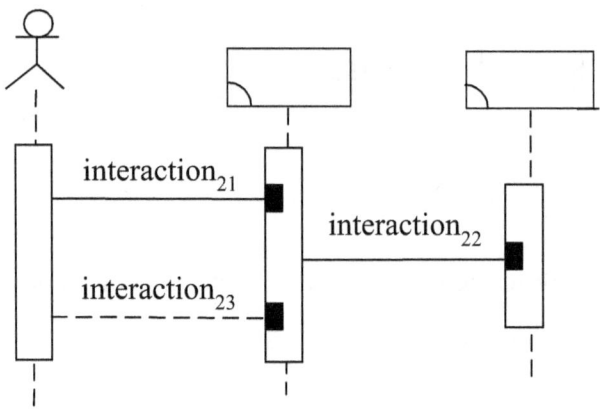

We formally describe this systems architecture as a 2-tuple IFDTUPLE = <IFD_1, IFD_2>.

Or we formally describe it as a 2-tuple IFDTUPLE = <($interaction_{11}$, $interaction_{12}$), ($interaction_{21}$, $interaction_{22}$, $interaction_{23}$)>.

PART II: EXECUTION OF THE SYSTEMS ARCHITECTURE

Sequential Execution of Interactions of an Interaction Flow Diagram

An interaction flow diagram may consist of many interactions. In this interaction flow diagram, each interaction will be sequentially executed.

The yth execution of the xth interaction flow diagram is defined as a sequence $IFDONCEEXECUTION_{xy}$ = (e_interaction$_{xyz}$) $_{z = 1\ to\ N}$, where "z" stands for the zth (z = 1 to N) interaction of the xth interaction flow diagram.

Examples of Sequential Execution of Interactions of an Interaction Flow Diagram

We suppose the example systems architecture consists of two interaction flow diagrams.

The first interaction flow diagram consists of two interactions.

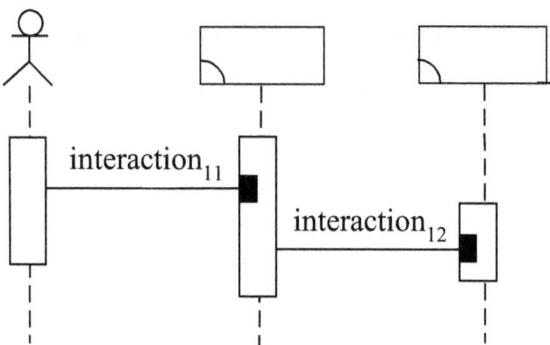

The yth execution of this 1st interaction flow diagram is defined as a sequence $IFDONCEEXECUTION_{1y}$ = $(e\_interaction_{1y1}, e\_interaction_{1y2})$.

The second interaction flow diagram consists of three interactions.

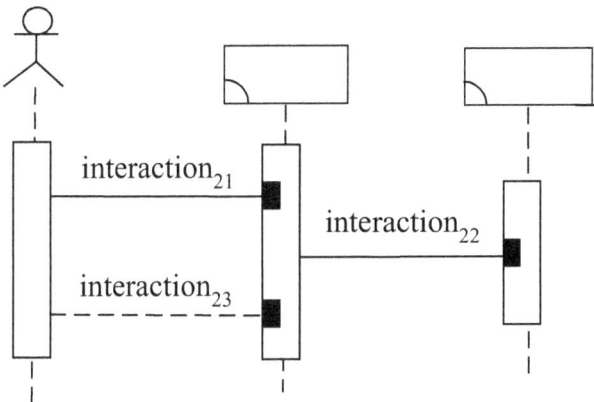

The yth execution of this 2nd interaction flow diagram is defined as a sequence $IFDONCEEXECUTION_{2y}$ = ($e\_interaction_{2y1}$, $e\_interaction_{2y2}$, $e\_interaction_{2y3}$).

Infinite Executions of an Interaction Flow Diagram

An interaction flow diagram may be executed an infinite number of times.

The infinite execution of the xth interaction flow diagram is defined as a sequence IFDINFINITEEXECUTION$_x$ = (e_interaction$_{xyz}$) $_{y = 1\ to\ \infty\ and\ z = 1\ to\ N}$, where "y" stands for the yth (y = 1 to ∞) execution of the xth interaction flow diagram; "z" stands for the zth (z = 1 to N) interaction of the xth interaction flow diagram.

Examples of Infinite Executions of an Interaction Flow Diagram

We suppose the example systems architecture consists of two interaction flow diagrams.

The first interaction flow diagram consists of two interactions.

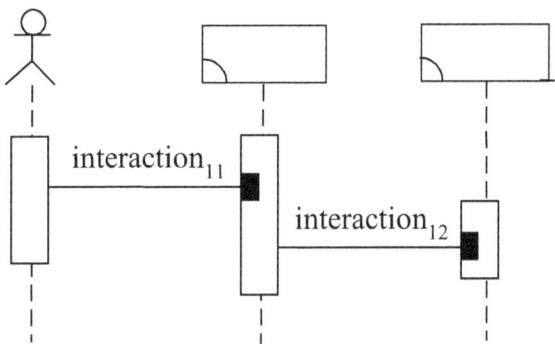

The infinite execution of this 1st interaction flow diagram is defined as a sequence $IFDINFINITEEXECUTION_1$ = ($e\_interaction_{111}$, $e\_interaction_{112}$, $e\_interaction_{121}$, $e\_interaction_{122}$, $e\_interaction_{131}$, $e\_interaction_{132}$, $e\_interaction_{141}$, $e\_interaction_{142}$, $e\_interaction_{151}$, $e\_interaction_{152}$, ..., $e\_interaction_{1\infty1}$, $e\_interaction_{1\infty2}$).

The second interaction flow diagram consists of three interactions.

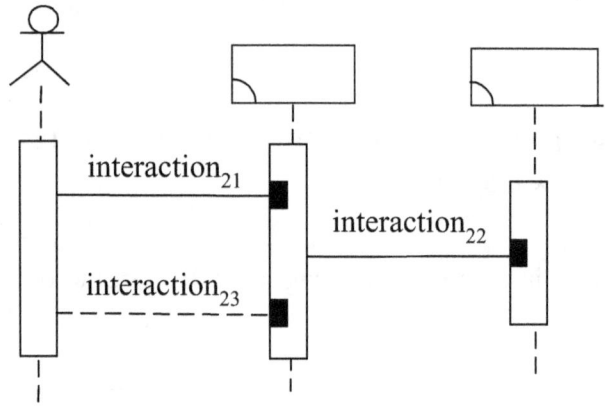

The infinite execution of this 2nd interaction flow diagram is defined as a sequence $IFDINFINITEEXECUTION_2$ = ($e\_interaction_{211}$, $e\_interaction_{212}$, $e\_interaction_{213}$, $e\_interaction_{221}$, $e\_interaction_{222}$, $e\_interaction_{223}$, $e\_interaction_{231}$, $e\_interaction_{232}$, $e\_interaction_{233}$, $e\_interaction_{241}$, $e\_interaction_{242}$, $e\_interaction_{243}$, $e\_interaction_{251}$, $e\_interaction_{252}$, $e\_interaction_{253}$,…, $e\_interaction_{2\infty1}$, $e\_interaction_{2\infty2}$, $e\_interaction_{2\infty3}$).

All Infinitely-Executed Individual Interaction Flow Diagrams Tend to Be Executed Concurrently

The systems architecture is a collection of all its individual interaction flow diagrams.

These infinitely-executed individual interaction flow diagrams are mutually independent of each other. They tend to be executed concurrently.

The concurrent execution of all infinitely-executed individual interaction flow diagrams means to define the execution of the systems architecture.

Execution of the Systems Architecture

The systems architecture may consist of many interaction flow diagrams. Each interaction flow diagram may be executed an infinite number of times. An interaction flow diagram may consist of many interactions.

The execution of the systems architecture is defined as a sequence SYSTEMSARCHITECTUREEXECUTION = (e_interaction$_{xyz}$) $_{x = 1 \text{ to } M \text{ and } y = 1 \text{ to } \infty \text{ and } z = 1 \text{ to } N}$, where "x" stands for the xth (x = 1 to M) interaction flow diagram of this systems architecture; "y" stands for the yth (y = 1 to ∞) execution of the xth interaction flow diagram; "z" stands for the zth (z = 1 to N) interaction of the xth interaction flow diagram.

Examples of Execution of the Systems Architecture

We suppose the example systems architecture consists of two interaction flow diagrams. The first interaction flow diagram consists of two interactions.

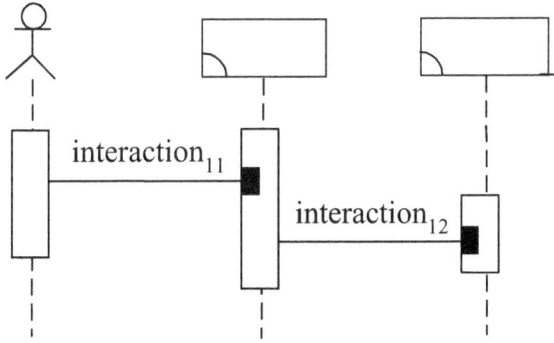

The second interaction flow diagram consists of three interactions.

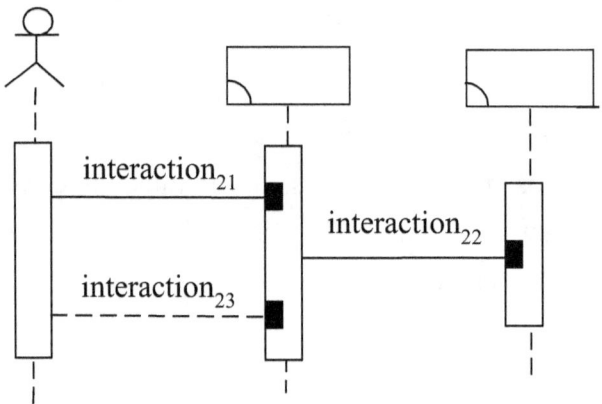

There are three (or more) possible executions of this systems architecture.

SYSTEMSARCHITECTUREEXECUTION^{SQM} =
(e_interaction$_{111}$, e_interaction$_{112}$, e_interaction$_{121}$, e_interaction$_{122}$, e_interaction$_{131}$, e_interaction$_{132}$, e_interaction$_{211}$, e_interaction$_{212}$, e_interaction$_{213}$, e_interaction$_{141}$, e_interaction$_{142}$, e_interaction$_{221}$, e_interaction$_{222}$, e_interaction$_{223}$,...) is the first possible execution of this systems architecture.

SYSTEMSARCHITECTUREEXECUTION^{MQM} $=$ ($e\_interaction_{111}$, $e\_interaction_{211}$, $e\_interaction_{212}$, $e\_interaction_{213}$, $e\_interaction_{112}$, $e\_interaction_{121}$, $e\_interaction_{122}$, $e\_interaction_{131}$, $e\_interaction_{132}$, $e\_interaction_{141}$, $e\_interaction_{142}$, $e\_interaction_{221}$, $e\_interaction_{222}$, $e\_interaction_{223}$,...) is the second possible execution of this systems architecture.

SYSTEMSARCHITECTUREEXECUTION^{IQM} $=$ ($e\_interaction_{111}$, $e\_interaction_{121}$, $e\_interaction_{131}$, $e\_interaction_{112}$, $e\_interaction_{122}$, $e\_interaction_{132}$, $e\_interaction_{211}$, $e\_interaction_{212}$, $e\_interaction_{213}$, $e\_interaction_{141}$, $e\_interaction_{142}$, $e\_interaction_{221}$, $e\_interaction_{222}$, $e\_interaction_{223}$,...) is third possible execution of this systems architecture.

.

Interactions Scheduling Defines the Execution of the Systems Architecture

The execution of the systems architecture is described by the scheduling of all interactions.

Different scheduling of all interactions will generate different sequences for executing the systems architecture.

There are at least three models (i.e. SQM, MQM, IQM) of interaction scheduling for the execution of the systems architecture. We will elaborate on their details in the following chapters.

Fairness of Interactions Scheduling

No matter what model we will use for scheduling of interactions, fairness is an important feature that any interactions scheduling algorithm must possess.

Fairness guarantees that every executable interaction will and shall eventually be executed.

PART III: SINGLE-QUEUE MODEL FOR INTERACTIONS SCHEDULING

Method of Single-Queue Model

Single-queue model (SQM) adopts a single queue interactions scheduling algorithm. There exists only one queue in this single-queue model.

Whenever the yth execution of the xth interaction flow diagram, IFD_x, is ready, its to-be-executed interactions such as e_interaction$_{xy1}$, e_interaction$_{xy2}$,..., e_interaction$_{xyN}$ will be one-by-one added at the end of the SYSTEMSARCHITECTUREEXECUTION_QUEUE queue. The *Ready_Head* is a pointer variable which points to the interaction at the head of the queue.

48

Ready_Head

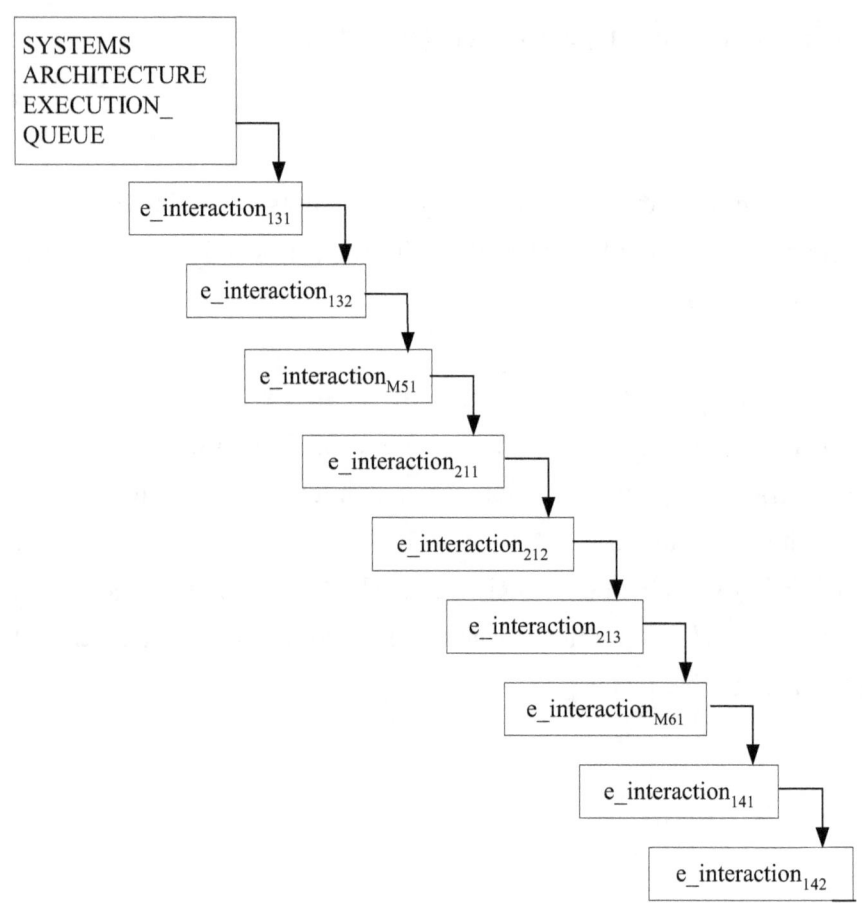

Given the queue structure just described, the SQM scheduling algorithm is simple. It just picks the to-be-executed interaction at the head of that queue. If the queue is empty, then the idle routine will be executed.

Features of Single-Queue Model

The sequence generated by the single-queue model for interactions scheduling algorithm possesses the following characteristics:

(1) It is a fair scheduling. Every executable interaction will and shall eventually be executed.

(2) If $e < f$, then the e_interaction$_{xye}$ must be executed before the e_interaction$_{xyf}$.

(3) If $a < b$, then the e_interaction$_{xae}$ must be executed before the e_interaction$_{xbf}$.

(4) If $e < f$, then only those e_interaction$_{xyg}$ which satisfy the "$e < g < f$" condition are allowed to be executed before the e_interaction$_{xyf}$ and after the e_interaction$_{xye}$.

PART IV: MULTI-QUEUE MODEL

FOR INTERACTIONS

SCHEDULING

Method of Multi-Queue Model

Multi-queue model (MQM) adopts a multiple queues interactions scheduling algorithm. The number of interaction flow diagrams will be used as the number of queues. All queues are treated equally, without any preference.

Whenever the yth execution of the xth interaction flow diagram, IFD_x, is ready, its to-be-executed interactions such as e_interaction$_{xy1}$, e_interaction$_{xy2}$,..., e_interaction$_{xyN}$ will be one-by-one added at the end of the IFDINFINITEEXECUTION$_x$_QUEUE queue. That is, all to-be-executed interactions of the 1st interaction flow diagram are queued in the IFDINFINITEEXECUTION$_1$_QUEUE queue; all to-be-executed interactions of the 2nd interaction flow diagram are queued in the IFDINFINITEEXECUTION$_2$_QUEUE queue;...; all to-be-executed interactions of the Mth interaction flow diagram are queued in the IFDINFINITEEXECUTION$_M$_QUEUE queue. The array *Ready_Head* has one entry for each queue, with that entry pointing to the interaction at the head of the queue.

Ready_Head

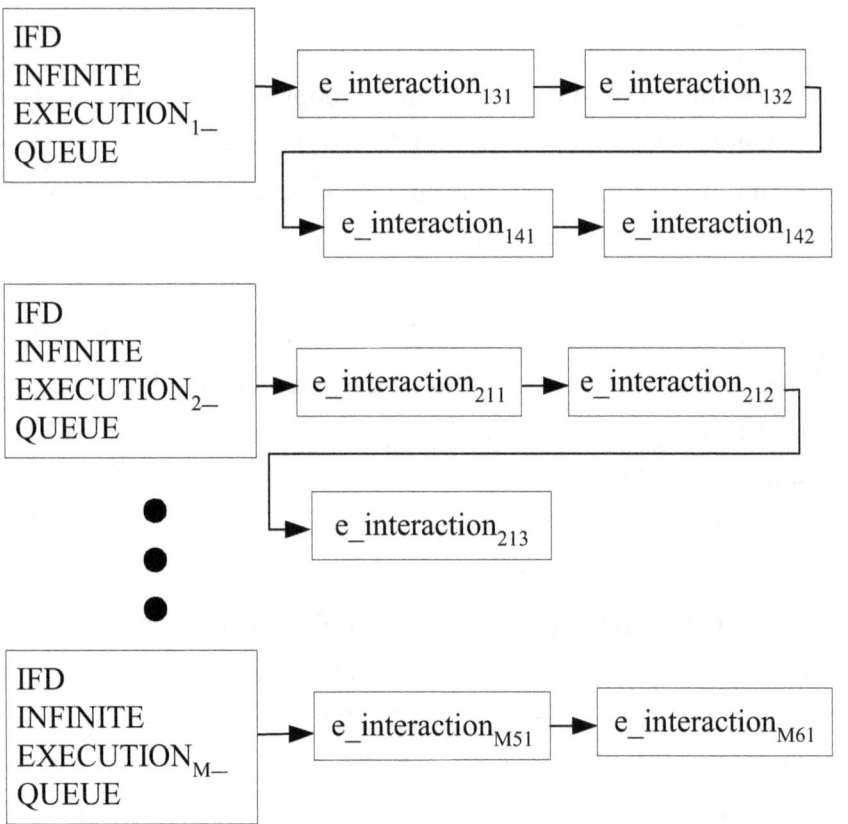

Given the queue structures just described, the MQM scheduling algorithm is simple. It just randomly chooses a queue that is not empty and picks the to-be-executed interaction at the head of that queue. If all the queues are empty, then the idle routine will be executed.

Features of Multi-Queue Model

The sequence generated by the multi-queue model for interactions scheduling algorithm possesses the following characteristics:

(1) It is a fair scheduling. Every executable interaction will and shall eventually be executed.

(2) If $e < f$, then the e_interaction$_{xye}$ must be executed before the e_interaction$_{xyf}$.

(3) If $a < b$, then the e_interaction$_{xae}$ must be executed before the e_interaction$_{xbf}$.

PART V: INFINITE-QUEUE
MODEL FOR INTERACTIONS
SCHEDULING

Method of Infinite-Queue Model

Infinite-queue model (IQM) adopts an infinite queues interactions scheduling algorithm. The number of queues is infinite. All queues are treated equally, without any preference.

Whenever the yth execution of the xth interaction flow diagram, IFD_x, is ready, a completely new queue, $IFDONCEEXECUTION_{xy}\_QUEUE$, will be prepared and all to-be-executed interactions of this yth execution such as $e\_interaction_{xy1}$, $e\_interaction_{xy2}$,…, $e\_interaction_{xyN}$ will be one-by-one added at the end of the $IFDONCEEXECUTION_{xy}\_QUEUE$ queue. Since there are infinite executions of an interaction flow diagram, there will be an infinite number of new queues.

The array *Ready_Head* has one entry for each queue, with that entry pointing to the interaction at the head of the queue.

Ready_Head

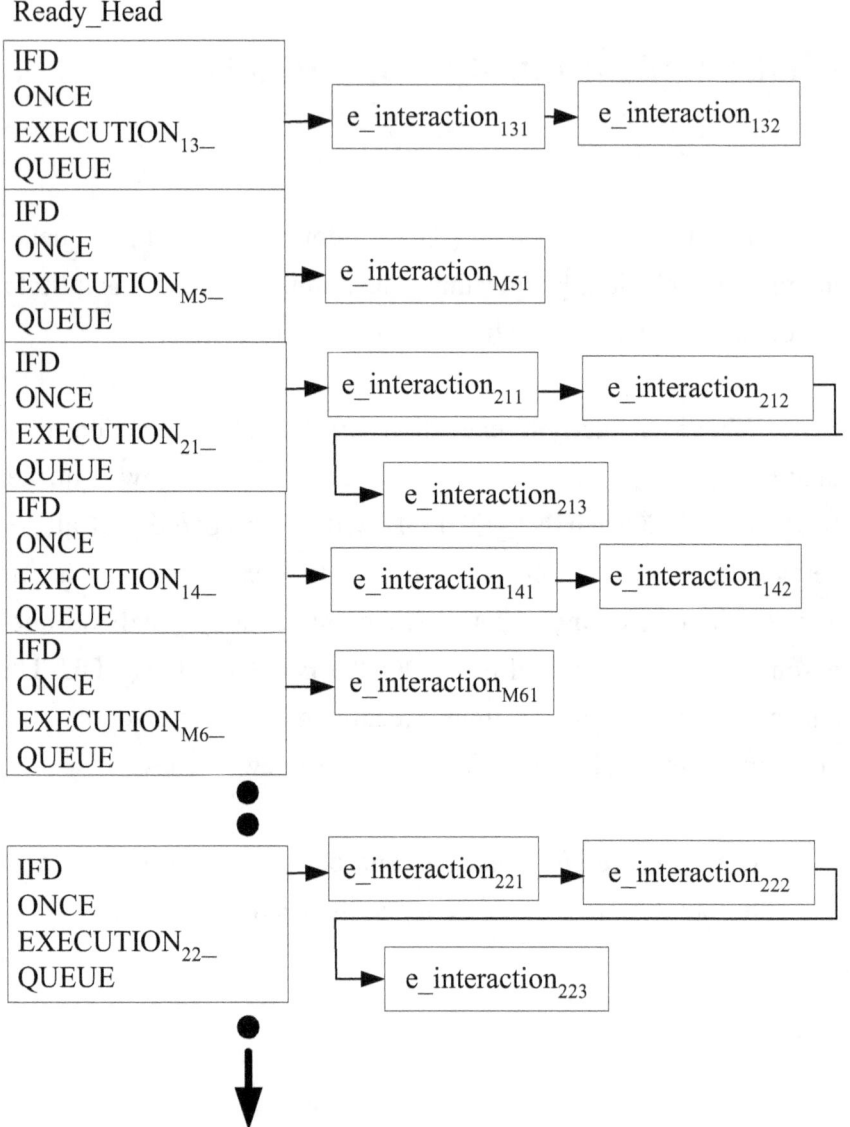

Given the queue structures just described, the IQM scheduling algorithm is simple. It just randomly chooses a queue that is not empty and picks the to-be-executed interaction at the head of that queue. If all the queues are empty, then the idle routine will be executed.

Features of Infinite-Queue Model

The sequence generated by the infinite-queue model for interactions scheduling algorithm possesses the following characteristics:

(1) It is a fair scheduling. Every executable interaction will and shall eventually be executed.

(2) If $e < f$, then the e_interaction$_{xye}$ must be executed before the e_interaction$_{xyf}$.

PART VI: VISIBILITY OF
VARIABLES

Scope of Name Binding

The scope of a name binding – an association of a name to an entity, such as a variable – is the part of the execution of the systems architecture.

The scope of a name binding is also known as the visibility of a variable.

The term "scope" is also used to refer to the set of all variables that are visible or names that are valid within a portion of the systems architecture.

Levels of Scope

There are eight levels of scope in the execution of the systems architecture.

Varied-xyz scope is also known as the global scope. A varied-xyz variable is a variable with varied-xyz scope, meaning that it is visible throughout all executed interactions of the systems architecture.

Varied-yz scope is also known as the fixed-x scope. A varied-yz variable is a variable with varied-yz scope, meaning that it is visible throughout fixed-x executed interactions of the systems architecture.

Varied-xy scope is also known as the fixed-z scope. A varied-xy variable is a variable with varied-xy scope, meaning that it is visible throughout fixed-z executed interactions of the systems architecture.

Varied-xz scope is also known as the fixed-y scope. A varied-xz variable is a variable with varied-xz scope, meaning that it is visible throughout fixed-y executed interactions of the systems architecture.

Fixed-xz scope is also known as the varied-y scope. A fixed-xz variable is a variable with fixed-xz scope, meaning that it is visible throughout varied-y executed interactions of the systems architecture.

Fixed-xy scope is also known as the varied-z scope. A fixed-xy variable is a variable with fixed-xy scope, meaning that it is visible throughout varied-z executed interactions of the systems architecture.

Fixed-yz scope is also known as the varied-x scope. A fixed-yz variable is a variable with fixed-yz scope, meaning that it is visible throughout varied-x executed interactions of the systems architecture.

Fixed-xyz scope is also known as the local scope. A fixed-xyz variable is a variable with fixed-xyz scope, meaning that it is only visible within each executed interaction of the systems architecture.

Set of all Executed Interactions of the Systems Architecture

An executed interaction of the systems architecture is defined as an element EXECUTEDINTERACTION = $e\_interaction_{xyz}$ where "x" stands for the xth interaction flow diagram of this systems architecture; "y" stands for the yth execution of the xth interaction flow diagram; "z" stands for the zth interaction of the xth interaction flow diagram.

The set of all executed interactions of the systems architecture is defined as a set SETOFALLEXECUTEDINTERACTIONS = $\{e\_interaction_{xyz}\}$ $_{x\,=\,1\,to\,M\,and\,y\,=\,1\,to\,\infty\,and\,z\,=\,1\,to\,N}$, where "x" stands for the xth (x = 1 to M) interaction flow diagram of this systems architecture; "y" stands for the yth (y = 1 to ∞) execution of the xth interaction flow diagram; "z" stands for the zth (z = 1 to N) interaction of the xth interaction flow diagram.

Varied-xyz Scope

A declaration has varied-xyz scope if it has effect throughout the set of {e_interaction$_{xyz}$} $_{x\,=\,1\ to\ M\ and\ y\,=\,1\ to\ \infty\ and\ z\,=\,1\ to\ N}$, where "x" stands for the xth (x = 1 to M) interaction flow diagram of this systems architecture; "y" stands for the yth (y = 1 to ∞) execution of the xth interaction flow diagram; "z" stands for the zth (z = 1 to N) interaction of the xth interaction flow diagram.

Varied-xyz scope is also known as the global scope. A varied-xyz variable is a variable with varied-xyz scope, meaning that it is visible (hence accessible) throughout all executed interactions of the systems architecture.

Varied-yz Scope

A declaration has varied-yz scope if it has effect throughout the set of $\{e\_interaction_{xyz}\}$ $_{x\,=\,x\,\text{to}\,x\,\text{and}\,y\,=\,1\,\text{to}\,\infty\,\text{and}\,z\,=\,1\,\text{to}\,N}$, where "x" stands for the xth (x = x to x) interaction flow diagram of this systems architecture; "y" stands for the yth (y = 1 to ∞) execution of the xth interaction flow diagram; "z" stands for the zth (z = 1 to N) interaction of the xth interaction flow diagram.

Varied-yz scope is also known as the fixed-x scope. A varied-yz variable is a variable with varied-yz scope, meaning that it is visible (hence accessible) throughout fixed-x executed interactions of the systems architecture.

Varied-xy Scope

A declaration has varied-xy scope if it has effect throughout the set of $\{e\_interaction_{xyz}\}$ $_{x\,=\,1\text{ to M and }y\,=\,1\text{ to }\infty\text{ and }z\,=\,z\text{ to z}}$, where "x" stands for the xth (x = 1 to M) interaction flow diagram of this systems architecture; "y" stands for the yth (y = 1 to ∞) execution of the xth interaction flow diagram; "z" stands for the zth (z = z to z) interaction of the xth interaction flow diagram.

Varied-xy scope is also known as the fixed-z scope. A varied-xy variable is a variable with varied-xy scope, meaning that it is visible (hence accessible) throughout fixed-z executed interactions of the systems architecture.

Varied-xz Scope

A declaration has varied-xz scope if it has effect throughout the set of $\{e\_interaction_{xyz}\}$ $_{x = 1 \text{ to } M \text{ and } y = y \text{ to } y \text{ and } z = 1 \text{ to } N,}$ where "x" stands for the xth (x = 1 to M) interaction flow diagram of this systems architecture; "y" stands for the yth (y = y to y) execution of the xth interaction flow diagram; "z" stands for the zth (z = 1 to N) interaction of the xth interaction flow diagram.

Varied-xz scope is also known as the fixed-y scope. A varied-xz variable is a variable with varied-xz scope, meaning that it is visible (hence accessible) throughout fixed-y executed interactions of the systems architecture.

Fixed-xz Scope

A declaration has fixed-xz scope if it has effect throughout the set of $\{e\_interaction_{xyz}\}$ $_{x\,=\,x\,\text{to}\,x\,\text{and}\,y\,=1\,\text{to}\,\infty\,\text{and}\,z\,=\,z\,\text{to}\,z}$, where "x" stands for the xth (x = x to x) interaction flow diagram of this systems architecture; "y" stands for the yth (y = 1 to ∞) execution of the xth interaction flow diagram; "z" stands for the zth (z = z to z) interaction of the xth interaction flow diagram.

Fixed-xz scope is also known as the varied-y scope. A fixed-xz variable is a variable with fixed-xz scope, meaning that it is visible (hence accessible) throughout varied-y executed interactions of the systems architecture.

Fixed-xy Scope

A declaration has fixed-xy scope if it has effect throughout the set of $\{e\_interaction_{xyz}\}$ x = x to x and y = y to y and z = 1 to N, where "x" stands for the xth (x = x to x) interaction flow diagram of this systems architecture; "y" stands for the yth (y = y to y) execution of the xth interaction flow diagram; "z" stands for the zth (z = 1 to N) interaction of the xth interaction flow diagram.

Fixed-xy scope is also known as the varied-z scope. A fixed-xy variable is a variable with fixed-xy scope, meaning that it is visible (hence accessible) throughout varied-z executed interactions of the systems architecture.

Fixed-yz Scope

A declaration has fixed-yz scope if it has effect throughout the set of $\{e\_interaction_{xyz}\}$ $_{x\,=\,1\,to\,M\,and\,y\,=\,y\,to\,y\,and\,z\,=\,z\,to\,z}$, where "x" stands for the xth (x = 1 to M) interaction flow diagram of this systems architecture; "y" stands for the yth (y = y to y) execution of the xth interaction flow diagram; "z" stands for the zth (z = z to z) interaction of the xth interaction flow diagram.

Fixed-yz scope is also known as the varied-x scope. A fixed-yz variable is a variable with fixed-yz scope, meaning that it is visible (hence accessible) throughout varied-x executed interactions of the systems architecture.

Fixed-xyz Scope

A declaration has fixed-xyz scope if it has effect throughout the set of {e_interaction$_{xyz}$} $_{x = x \text{ to } x}$ and $_{y = y \text{ to } y}$ and $_{z = z \text{ to } z}$, where "x" stands for the xth (x = x to x) interaction flow diagram of this systems architecture; "y" stands for the yth (y = y to y) execution of the xth interaction flow diagram; "z" stands for the zth (z = z to z) interaction of the xth interaction flow diagram.

Fixed-xyz scope is also known as the local scope. A fixed-xyz variable is a variable with fixed-xyz scope, meaning that it is only visible (hence accessible) within each executed interaction of the systems architecture.

Examples of Scope of Name Binding

We suppose the example systems architecture consists of two interaction flow diagrams. The first interaction flow diagram consists of two interactions.

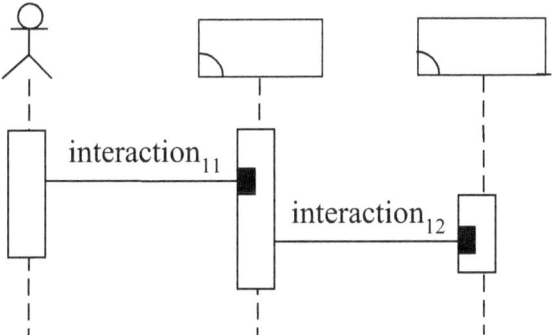

The second interaction flow diagram consists of three interactions.

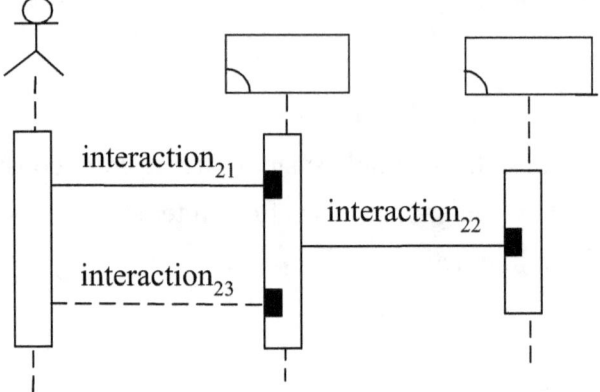

For this systems architecture, a declaration has varied-xyz (global) scope if it has effect throughout the set of

$\{$e_interaction$_{111}$, e_interaction$_{112}$, e_interaction$_{211}$, e_interaction$_{212}$, e_interaction$_{213}$, e_interaction$_{121}$, e_interaction$_{122}$, e_interaction$_{221}$, e_interaction$_{222}$, e_interaction$_{223}$, e_interaction$_{131}$, e_interaction$_{132}$, e_interaction$_{231}$, e_interaction$_{232}$, e_interaction$_{233}$, e_interaction$_{141}$, e_interaction$_{142}$, e_interaction$_{241}$, e_interaction$_{242}$, e_interaction$_{243}$, e_interaction$_{151}$, e_interaction$_{152}$, e_interaction$_{251}$, e_interaction$_{252}$, e_interaction$_{253}$,..., e_interaction$_{1\infty1}$, e_interaction$_{1\infty2}$, e_interaction$_{2\infty1}$, e_interaction$_{2\infty2}$, e_interaction$_{2\infty3}\}$

For this systems architecture, a declaration has varied-yz (fixed-x) scope if it has effect throughout the set of

$\{$e_interaction$_{111}$, e_interaction$_{121}$, e_interaction$_{131}$, e_interaction$_{141}$, e_interaction$_{151}$,..., e_interaction$_{1\infty1}$, e_interaction$_{112}$, e_interaction$_{122}$, e_interaction$_{132}$, e_interaction$_{142}$, e_interaction$_{152}$,..., e_interaction$_{1\infty2}\}$

OR

$\{$e_interaction$_{211}$, e_interaction$_{221}$, e_interaction$_{231}$, e_interaction$_{241}$, e_interaction$_{251}$,..., e_interaction$_{2\infty1}$, e_interaction$_{212}$, e_interaction$_{222}$, e_interaction$_{232}$, e_interaction$_{242}$, e_interaction$_{252}$,..., e_interaction$_{2\infty2}$,..., e_interaction$_{213}$, e_interaction$_{223}$, e_interaction$_{233}$, e_interaction$_{243}$, e_interaction$_{253}$,..., e_interaction$_{2\infty3}\}$

For this systems architecture, a declaration has varied-xy (fixed-z) scope if it has effect throughout the set of

$\{$e_interaction$_{111}$, e_interaction$_{121}$, e_interaction$_{131}$, e_interaction$_{141}$, e_interaction$_{151}$,..., e_interaction$_{1\infty1}$, e_interaction$_{211}$, e_interaction$_{221}$, e_interaction$_{231}$, e_interaction$_{241}$, e_interaction$_{251}$,..., e_interaction$_{2\infty1}\}$

OR

$\{$e_interaction$_{112}$, e_interaction$_{122}$, e_interaction$_{132}$, e_interaction$_{142}$, e_interaction$_{152}$,..., e_interaction$_{1\infty2}$, e_interaction$_{212}$, e_interaction$_{222}$, e_interaction$_{232}$, e_interaction$_{242}$, e_interaction$_{252}$,..., e_interaction$_{2\infty2}\}$

OR

$\{$e_interaction$_{213}$, e_interaction$_{223}$, e_interaction$_{233}$, e_interaction$_{243}$, e_interaction$_{253}$,..., e_interaction$_{2\infty3}\}$

82

For this systems architecture, a declaration has varied-xz (fixed-y) scope if it has effect throughout the set of

$\{e\_interaction_{111}, e\_interaction_{112}, e\_interaction_{211}, e\_interaction_{212}, e\_interaction_{213}\}$

OR

$\{e\_interaction_{121}, e\_interaction_{122}, e\_interaction_{221}, e\_interaction_{222}, e\_interaction_{223}\}$

OR

$\{e\_interaction_{131}, e\_interaction_{132}, e\_interaction_{231}, e\_interaction_{232}, e\_interaction_{233}\}$

OR

$\{e\_interaction_{141}, e\_interaction_{142}, e\_interaction_{241}, e\_interaction_{242}, e\_interaction_{243}\}$

OR

$\{e\_interaction_{151}, e\_interaction_{152}, e\_interaction_{251}, e\_interaction_{252}, e\_interaction_{253}\}$

. . . .

OR

$\{e\_interaction_{1\infty1}, e\_interaction_{1\infty2}, e\_interaction_{2\infty1}, e\_interaction_{2\infty2}, e\_interaction_{2\infty3}\}$

For this systems architecture, a declaration has fixed-xz (varied-y) scope if it has effect throughout the set of

$\{e\_interaction_{111}, e\_interaction_{121}, e\_interaction_{131}, e\_interaction_{141}, e\_interaction_{151},..., e\_interaction_{1\infty 1}\}$

OR

$\{e\_interaction_{112}, e\_interaction_{122}, e\_interaction_{132}, e\_interaction_{142}, e\_interaction_{152},..., e\_interaction_{1\infty 2}\}$

OR

$\{e\_interaction_{211}, e\_interaction_{221}, e\_interaction_{231}, e\_interaction_{241}, e\_interaction_{251},..., e\_interaction_{2\infty 1}\}$

OR

$\{e\_interaction_{212}, e\_interaction_{222}, e\_interaction_{232}, e\_interaction_{242}, e\_interaction_{252},..., e\_interaction_{2\infty 2}\}$

OR

$\{e\_interaction_{213}, e\_interaction_{223}, e\_interaction_{233}, e\_interaction_{243}, e\_interaction_{253},..., e\_interaction_{2\infty 3}\}$

For this systems architecture, a declaration has fixed-xy (varied-z) scope if it has effect throughout the set of

$\{e\_interaction_{111}, e\_interaction_{112}\}$

OR

$\{e\_interaction_{211}, e\_interaction_{212}, e\_interaction_{213}\}$

OR

$\{e\_interaction_{121}, e\_interaction_{122}\}$

OR

$\{e\_interaction_{221}, e\_interaction_{222}, e\_interaction_{223}\}$

OR

$\{e\_interaction_{131}, e\_interaction_{132}\}$

OR

$\{e\_interaction_{231}, e\_interaction_{232}, e\_interaction_{233}\}$

. . . .

OR

$\{e\_interaction_{1\infty1}, e\_interaction_{1\infty2}\}$

OR

$\{e\_interaction_{2\infty1}, e\_interaction_{2\infty2}, e\_interaction_{2\infty3}\}$

For this systems architecture, a declaration has fixed-yz (varied-x) scope if it has effect throughout the set of

$\{e\_interaction_{111}, \quad e\_interaction_{211}\}$ OR $\{e\_interaction_{112}, e\_interaction_{212}\}$ OR $\{e\_interaction_{213}\}$

OR

$\{e\_interaction_{121}, \quad e\_interaction_{221}\}$ OR $\{e\_interaction_{122}, e\_interaction_{222}\}$ OR $\{e\_interaction_{223}\}$

OR

$\{e\_interaction_{131}, \quad e\_interaction_{231}\}$ OR $\{e\_interaction_{132}, e\_interaction_{232}\}$ OR $\{e\_interaction_{233}\}$

OR

$\{e\_interaction_{141}, \quad e\_interaction_{241}\}$ OR $\{e\_interaction_{142}, e\_interaction_{242}\}$ OR $\{e\_interaction_{243}\}$

OR

$\{e\_interaction_{151}, \quad e\_interaction_{251}\}$ OR $\{e\_interaction_{152}, e\_interaction_{252}\}$ OR $\{e\_interaction_{253}\}$

. . . .

OR

$\{e\_interaction_{1\infty1}, \quad e\_interaction_{2\infty1}\}$ OR $\{e\_interaction_{1\infty2}, e\_interaction_{2\infty2}\}$ OR $\{e\_interaction_{2\infty3}\}$

For this systems architecture, a declaration has fixed-xyz (local) scope if it has effect throughout the set of

$\{e\_interaction_{111}\}$ OR $\{e\_interaction_{112}\}$ OR $\{e\_interaction_{211}\}$ OR $\{e\_interaction_{212}\}$ OR $\{e\_interaction_{213}\}$

OR

$\{e\_interaction_{121}\}$ OR $\{e\_interaction_{122}\}$ OR $\{e\_interaction_{221}\}$ OR $\{e\_interaction_{222}\}$ OR $\{e\_interaction_{223}\}$

OR

$\{e\_interaction_{131}\}$ OR $\{e\_interaction_{132}\}$ OR $\{e\_interaction_{231}\}$ OR $\{e\_interaction_{232}\}$ OR $\{e\_interaction_{233}\}$

OR

$\{e\_interaction_{141}\}$ OR $\{e\_interaction_{142}\}$ OR $\{e\_interaction_{241}\}$ OR $\{e\_interaction_{242}\}$ OR $\{e\_interaction_{243}\}$

OR

$\{e\_interaction_{151}\}$ OR $\{e\_interaction_{152}\}$ OR $\{e\_interaction_{251}\}$ OR $\{e\_interaction_{252}\}$ OR $\{e\_interaction_{253}\}$

. . . .

OR

$\{e\_interaction_{1\infty1}\}$ OR $\{e\_interaction_{1\infty2}\}$ OR $\{e\_interaction_{2\infty1}\}$ OR $\{e\_interaction_{2\infty2}\}$ OR $\{e\_interaction_{2\infty3}\}$

www.ingramcontent.com/pod-product-compliance
Lightning Source LLC
Chambersburg PA
CBHW070840180526
45168CB00002B/904